M000251750

FIESELER Fi 156 STORCH

Heinz J. Nowarra

Schiffer Military History
Atglen, PA

Sources:
Nowarra-Kens, Die deutschen Flugzeuge 1933-45
Dipl.Ing. K. Kössler, LBA Braunschweig

Picture Credits:
Nowarra
Petrick, Berlin
Kössler, Braunschweig
Krzyzan, Zielona Gora
Zazvonil, Prague
Postma, Hoofddorp

Translated from the German by Don Cox.
Cover artwork by Steve Ferguson.

Copyright © 1997 by Schiffer Publishing Ltd.

All rights reserved. No part of this work may be reproduced or used in any forms or by any means – graphic, electronic or mechanical, including photocopying or information storage and retrieval systems – without written permission from the copyright holder.

Printed in China.
ISBN: 0-7643-0299-X

This book was originally published under the title,
Fieseler 156 "Storch",
by Podzun-Pallas Verlag.

We are interested in hearing from authors with book ideas on related topics.

Published by Schiffer Publishing Ltd.
4880 Lower Valley Road
Atglen, PA 19310
Phone: (610) 593-1777
FAX: (610) 593-2002
E-mail: Schifferbk@aol.com.
Please write for a free catalog.
This book may be purchased from the publisher.
Please include $3.95 postage.
Try your bookstore first.

Fi 156 A-0, D-IJFN, Werknr. 612, August 1937.

Fieseler Fi 156 Storch

The concept of obtaining the greatest possible span between an aircraft's maximum and minimum speeds with a low-performing engine, thereby shortening landing and takeoff runs (now known as STOL short takeoff and landing), was an idea embraced back in 1930 by Professor Schmeidler of the Technische Hochschule in Breslau. Working in cooperation with Dipl.Ing. Neumann, he incorporated this concept into the design of various prototype models which achieved the sought-after characteristics by having their wing dimensions increased. Professor Winter of the Technische Hochschule in Braunschweig approached the problem from a different angle. He hoped to obtain the most favorable flight characteristics through the use of adjustable incidence angles for the wing in conjunction with high performance landing aids (i.e. flaps and slats). The result was the slow-flying "Zaunkönig", which achieved a cruising speed of 141 km/h and a landing speed of just 46 km/h using a 50 hp engine. The English were so fascinated by the airplane that they took it back with them, gave it the registration code of VX 190, and continued flying it long after the war was over. The efforts of Schmeidler, Neumann and Winter formed the basis for the development of the Fi 156.

In the spring of 1935 the Luftwaffe's Technisches Amt (LC) issued a requirement for an aircraft that would have extremely short takeoff and landing characteristics. The machine was to be utilized as a staff liaison aircraft for both ground and flying units as well as for directing artillery fire in other words, primarily for Army purposes.

Above right: Technische Hochschule Dresden, prototype of the Schmeidler-Jeschke SJ from 1930 with its adjustable wing chord. Powerplant: 45 hp BMW.

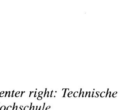

Center right: Technische Hochschule Braunschweig's Zaunkönig prior to its maiden flight.

Right: Professor Winter's Zaunkönig in England after 1945. With a maximum cruising speed of 141 km/h the aircraft's landing speed was just 46 km/h.

Above: Siebel(SFW) Si 201 V 1, D-IWHI.
Below: Focke-Wulf Fw 186 V 1, D-ISTQ.

Focke-Wulf, Fieseler, BFW (Messerschmitt) and Siebel all submitted proposals, designated Fw 186, Fi 156, Bf 163 and Si 201, respectively.

The Si 201 V 1 was about as unconventional as can be imagined: box-like rectangular cockpit employing a considerable amount of canopy glazing, observer sat fore and to the right, slender tubular boom to carry the tail surfaces. It employed a pusher-type engine/propeller combination above the wing trailing edge and, with its 240 hp Argus As 10 c engine, achieved a speed of 185 km/h. Generalluftzeugmeister Udet flew the aircraft, but pronounced its landing characteristics and ground handling qualities to be unsatisfactory.

Focke-Wulf, whose Professor Heinrich Focke was working on a German helicopter design following detailed studies of their license-built de la Cierva C 19 and C 30 autogyros, submitted the unusual Fw 186 autogyro as its entry. Fw 186 D-ISTQ was also rejected by Udet, since the aircraft would have been too temperamental for front-line service.

Messerschmitt (BFW) was building the Bf 163 for the requirement, but due to work on the Bf 109 wasn't able to meet the deadline. It was therefore proposed to LC that the initial preliminary work be transferred to someone else. The choice fell upon Weserflug's Dr. Adolf Rohrbach, who was most reluctant to continue with a project someone else had already begun.

Construction of the Bf 163 was carried out in the old Rohrbach facilities in the Kiautschoustrasse, Berlin North. It wasn't until early 1938 that Bf 163 V 1, D-IUCY, was ready for its first flight. In the interim, however, the decision had already been made in favor of the Fi 156. Two additional Bf 163s were never completed as a result.

The Fi 156 V 1 was assigned Werknummer (factory number) 601 and given the designation D-IBXY. However, no photo of this first airplane has ever been found. The Fi 156 V 2, Werknummer 602, D-IGLI, was probably under construction at the same time as the V 1 but, for reasons unknown, was finished earlier; D-IGLI was transferred to the Luftwaffe's Rechlin Test Center as early as 29 September 1936 whereas D-IBXY arrived there on 10 or 11 October 1936. Both machines differed from each other as follows:

- D-IGLI   Landing gear with continuous all-through axle, separated in the center, fixed slats over 2/3 of wingspan.
- D-IBXY   Landing gear as V 2, but fixed slats over 5/6 of wingspan.

During the course of testing the Fi 156 V 1 was damaged by the tow cable of a landing Ju 52, while the V 2's landing gear collapsed and had to be returned back to Fieseler in Kassel. Among other problems, the Rechlin engineers criticized both of these aircraft for:

Rudder forces being too pronounced, particularly with the horizontal stabilizer.

Insufficient horizontal stabilizer control for maintaining a stationary three-point stalled flight profile. The aircraft begins to shudder and noses over.

Extension of the landing flaps takes too long.

Landing gear too soft, all-through axle doesn't give adequate taxiing qualities in tall grass.

A positive feature was felt to be the aircraft's outstanding visibility in any flight attitude.

Above: Fieseler Fi 156 V 2, Werknr. 602, October 1936 in Rechlin.
Below: BFW's model of the Me 163 V 1, D-IUCY.

Fi 156 V 2 after its modification, seen in March 1939 on "Wehrmacht Day" in front of the Neue Wache on Berlin's Unter den Linden.

Fi 156 V 3, Werknummer 603, D-IGQE, served as a testbed for the installation of various radio sets the Fi 156 normally flew without radio. Later variants were equipped with the FuG VII, FuG 17 and FuG 14 radio sets. Fi 156 V 4 was probably used as an airframe testbed for researching airframe strength and vibration.

In 1937 the first pre-production series, the Fi 156 A-0, Werknummer 605 to 614, began rolling off the assembly lines, all of which were reserved for testing or for front-line trials. Werknummer 613, D-IKQD, flew in 1939 as a liaison aircraft with Lehrgeschwader 2 under the registration L2+039.

Based on experience at Rechlin, the first two Fi 156s were fundamentally modified in three specific areas: the slats were shortened, the continuous axle was dropped and the wings continued straight across with no dihedral. These changes were also incorporated on the A-0 series. There appears to have only been a limited number of A-1s produced, these being Werknummer 615 to 620. In the meantime, production had already started on the pre-production B-0 series, beginning with Werknummer 621 (registration WL-IHKV, formerly D-IHKV).

The Fi 156 V 2 was officially unveiled in its new form, no less in March of 1939 on "Wehrmacht Day" when it landed in Berlin's Unter den Linden between the State Opera House and the Neue Wache.

Above right and below: Fi 156 A-0, Werknr. 611 was registered on 15 July 1937 and entered flight testing with the code D-IDVS.

Fi 156 B-0, D-IKVN, Werknr. 625, was for years considered to be original Fi 156 V 1 due to the fact that in 1938 it was featured in the first pictures of an Fi 156 to be released.

An Fi 156 with its factory crew during the 1936 Olympic Games, seen in Berlin's Olympic Stadium.

Above right and right: D-IKVN was only cleared for flight trials in January 1938 and given over to the Luftwaffe in 1940 under the code DK+TV.

A B-1 series was apparently never built. Full series production only began with the C variants. The C-1 was generally comparable to the B-0, but unlike the C-2, C-3 and C-3 trop this version was not fitted with an MG 15 in a fisheye mount at the rear part of the canopy glazing. All C-series could be fitted with the "Spinne" cable-laying device as needed. The D-series (D-0, D-1 trop and D-2) could be easily recognized by their triangular window behind the standard canopy glazing. The various canopy glazing styles of the individual variants can be seen from the included drawings.

Since the differences between the various C- and D-series versions can generally only be discerned from within the aircraft itself, the photo captions do not specify the types beyond the general A, B, C and D classification. Internal variations among the individual versions are shown in the table at the back of the book.

Right: Variations in canopy glazing.
Below left: Fi 156 C-2 fuselage framework.
Below: Fi 156 C-1 fuselage framework.

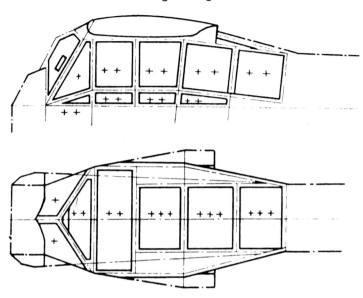

Kabinenverglasung C-1

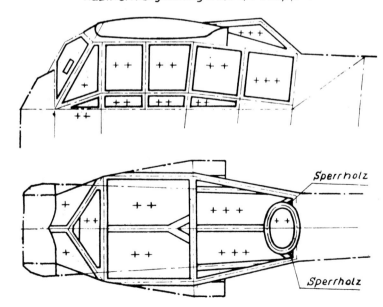

Kabinenverglasung C-2, C-3, C-3 trop, F-1

Sperrholz

Sperrholz

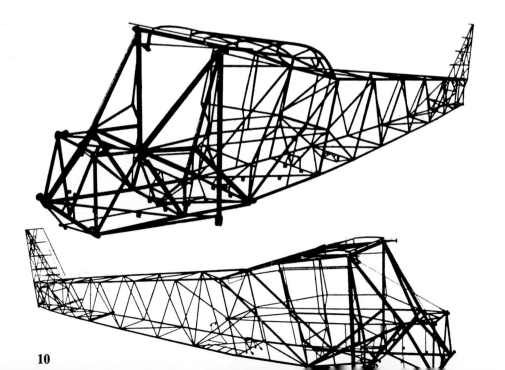

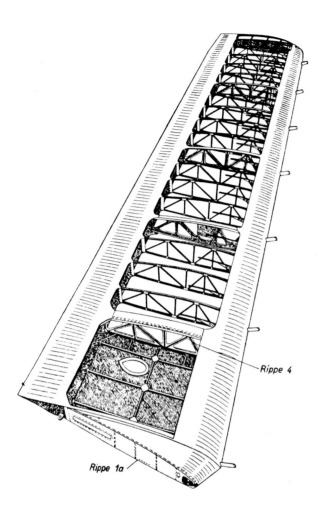

Rippe 4

Rippe 1a

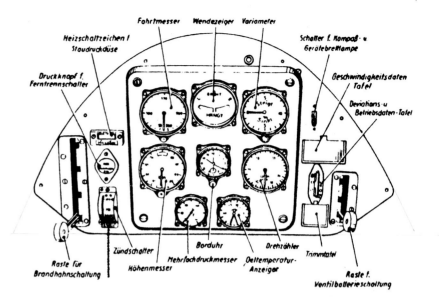

Heizschaltzeichen /
Staudruckdüse

Druckknopf f.
Ferntrennschalter

Fahrtmesser   Wendezeiger   Variometer

Schalter f. Kompaß- u.
Gerätebrettlampe

Geschwindigkeitsdaten
Tafel

Deviations- u.
Betriebsdaten-Tafel

Raste für
Brandhahnschaltung

Zündschalter

Höhenmesser

Mehrfachdruckmesser

Borduhr

Oeltemperatur-
Anzeiger

Drehzähler

Trimmtafel

Raste f.
Ventilbatterieschaltung

Above: Fi 156 wing construction.
Above right: Instrumentation layout.
Right: An Fi 156 A-0 rolls out of the assembly
hangar in Kassel.

Fi 156 C during acceptance flight. Flaps and control surface layout are clearly shown.

Above: Fi 156 B-0.
Below: Fi 156 B-0 in slow flight.

Above: An Italian military delegation inspects Fi 156 B-0 D-IFHR.
Below: Fi 156 B-0, D-IFBK, demonstrating its STOL capabilities.

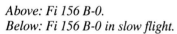

Fi 156 B-0, D-IFMR, during winter trials with auxiliary fuel tank and snow skis.

The Storch's ability to take off and land in the smallest confines was astounding; here one demonstrates these features in a barracks courtyard.

Fi 156 C, Werknr. 631, and two additional aircraft destined for Sweden.

A lineup of Fi 156 C models seen in Karlsbad during the annexation of Czechoslovakia.

The Storch also proved its abilities while in service with the Condor Legion.

Storchs of the Condor Legion's HQ unit.

The 27th aircraft of the Fi 156 C-2 series in flight. The fisheye mount for the MG 15 (LL 15) can clearly be seen.

This Fi 156 C landed in 1940 at the Place de la Concorde in Paris.

In the field, refuelling had to be accomplished by hand.

Crash landing somewhere on the Western Front, 1940.

An Fi 156 C-2 over France.

Above: Major Beck, Kommodore of JG 1, posing in front of the Geschwader's Storch.

Right: Fi 156 with the badge of 4/ (H)21, a Heeresfliegerstaffel (Army flying squadron).

Right: Fi 156 C-2 with LL 15.

Eastern Front, 2 October 1941: Generalmajor Freiherr von Richthofen, Kommodore of the VIII Fliegerkorps, seen with his Storch (a C-2 with FuG VII) visiting Panzer-Regiment 11, north of the taxiway east of Demidov at the beginning of the Battle of Vyazma. To his right is Oberst Koll, commander of Pz.Rgt. 11.

Fi 156 C, 17 September 1943, laying cable somewhere along the Eastern Front.

Fi 156 C of a Nah-Aufklärungs-Gruppe (NAG=Tactical Reconnaissance Group) somewhere on the Eastern Front.

Fi 156 C-2, summer of 1941 in the East.

Two Storchs at a field airstrip in the East, 1942.

Above:
Fi 156 C-2, TK+JE, Russia 1941.

Left:
This Fi 156 C served III Gruppe of Kampfgeschwader 4 as a liaison aircraft.

The Fi 156 P appeared in 1942, designed as a specialized aircraft for engaging partisan forces. This type differed from all other models by having racks for canisters holding 24 SD 2/XII fragmentation bombs. These were thick-walled 2 kg bombs with a diameter of 78 mm and a length of 303 mm, each having 0.225 kg explosive charge. In place of the bombs, smoke laying devices could be fitted for the purpose of obscuring important sites from high altitude Allied bomber formations.

The role which the Fi 156 played as a "flying observation post" for the Army's staff units has been well documented. General Rommel's North African Storch was particularly well known. Another Storch played a key military role albeit inadvertently on 12 September 1941, when the commander of the 11th Army, Generaloberst Ritter von Schobert, was forced to land in a Soviet minefield and was killed along with his pilot. His successor was the future Generalfeldmarschall von Manstein, who later conquered the Crimea with this same army.

It will be remembered that the Storch was involved in another important, although not militarily decisive, mission. On 12 September 1943 1 Kompanie of the Fallschirmjäger-Lehrbataillons under the command of Oberleutnant von Berlepsch landed with their transport gliders at the Campo Imperatore Hotel in Gran Sasso in the Abruzzi Mountains and there freed the imprisoned Mussolini. Hauptmann Gerlach then flew him out to freedom with an Fi 156 C-5 Storch, coded SJ+LL.

Above: Poland 1939. This Fi 156 C has been camouflaged, but still retains the earlier registration code of WL-1106.

Below: Fi 156 C in the East, 1942.

Above: Fi 156 P, 1942: hanging an SC 50 canister bomb.
Above right: Fi 156 P with smoke dispensers.
Right: Underfuselage bomb racks.

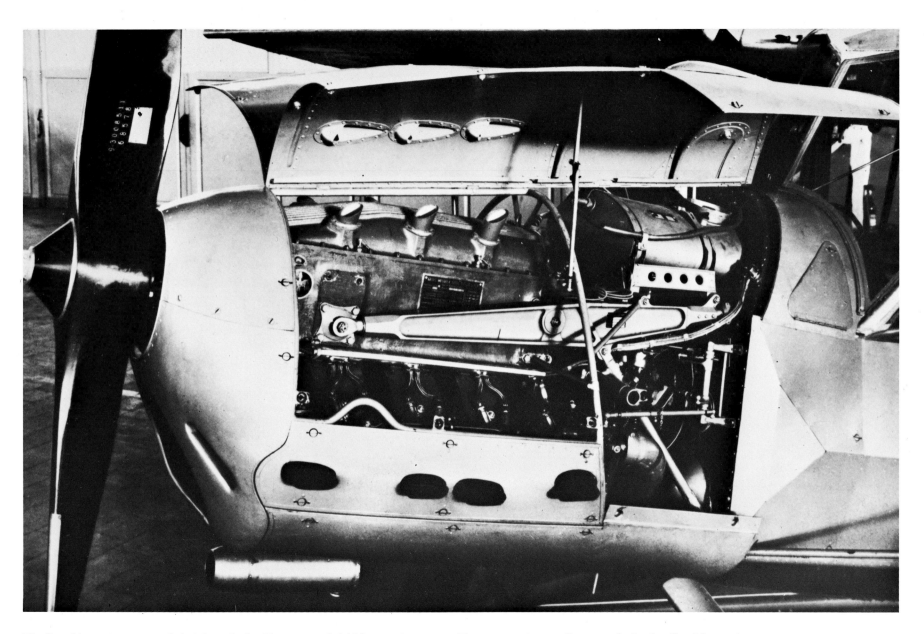

The Storch's engine: air-cooled eight-cylinder V engine with 240 hp starting power. The open engine cowling reveals the details of the engine support frame.

Fi 156 C in the spring of 1941 at Stagnone on Sicily.

A Storch over the Libyan desert.

The Storch was often called upon to rescue downed crews in Africa.

This was the Fi 156 C-5 (SJ+LL) with which Hptm. Gerlach flew the freed Mussolini out of Gran Sasso.

This Fi 156 C was left behind in Marmarica in the wake of Rommel's retreat and was flown for a time with British markings.

Russia 1941: Amazed, Russian children surround a Storch.

Above:
Field applied winter camouflage on a ski-equipped Fi 156.
Right:
The Storchs had to be continually kept clear of snow.

Winter of 1941/42 in the northern sector of the Eastern Front. The commanding general of XLI Panzerkorps uses a Storch from 4/(H)21 to visit a forward unit.

Fi 156 C takes off from a planked strip in the Ukraine, which has been laid down over the mud.

Fi 156 C of a NAG in the East, 1943.

Storchs were also used in the far north, in Finland and Norway.

Winter camouflage was generally applied within the units themselves.

This is the observer's view of the pilot's seat and instrument panel in an Fi 156.

Although the Storch was initially built exclusively by Fieseler, license building firms had to be brought into the program once Fieseler increasingly became involved in the license production of the Fw 190. From April 1942 on the firm of Morane-Saulnier in Le Puteaux began producing the Fi 156. Due to Fw 190 construction Fieseler was forced to drop all work on the manufacture of the Fi 156 from October 1943 onward. At this point the firm of Leichtbau Budweis in the Protectorate of Bohemia and Moravia (Czechoslovakia) was brought in to the production program, although it only built one Storch in 1943. In 1944, however, 72 were built there before production was transferred to the firm of Mraz in Chozen, also in Czechoslovakia. Mraz produced an additional 64 Fi 156s for the Germans. With 884 units built, 1943 was the best year for production of the Fi 156. Altogether 2,874 Fi 156s were built between 1936 and 1944.

In 1943 one or two prototypes of an improved Storch were built, the Fi 256. This machine was powered by the Argus As 10 P, carried a pilot and three to four crew. With an empty weight of 1,200 kg, takeoff weight was 1,680 kg. The aircraft had a range of 730 km. However, it never achieved production status.

Above: Fi 156 C with auxiliary fuel tank beneath the fuselage.
Right: Storch with MG (LL 15) and skis of a Luftwaffe subordinated Croatian squadron.

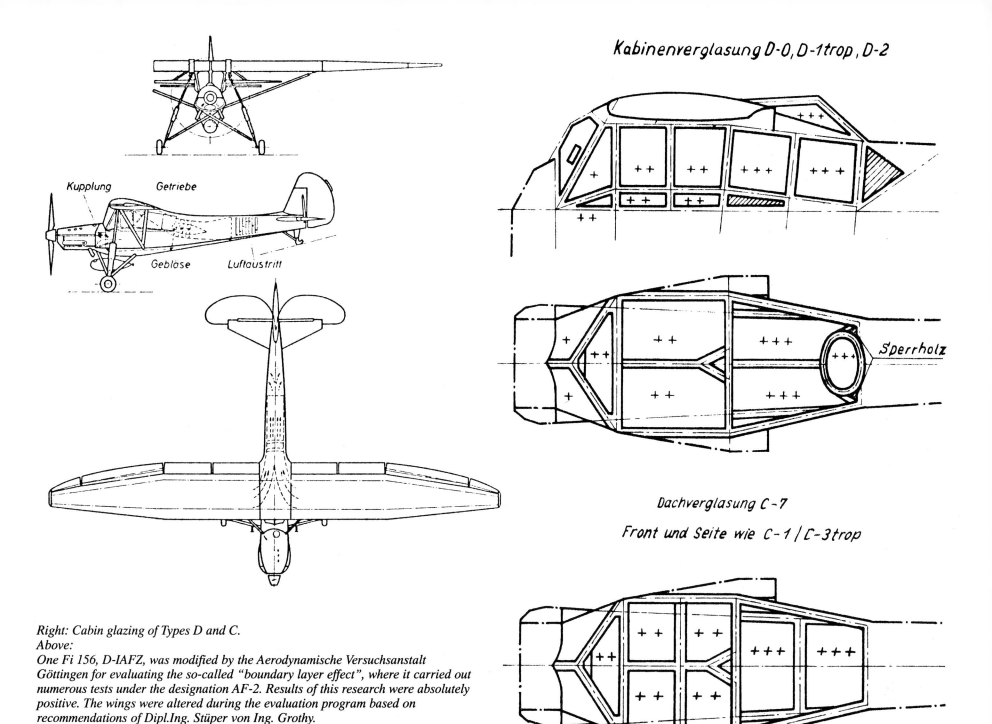

Kupplung Getriebe

Gebläse Luftaustritt

Kabinenverglasung D-0, D-1trop, D-2

Sperrholz

Dachverglasung C-7

Front und Seite wie C-1/C-3trop

Right: Cabin glazing of Types D and C.
Above:
One Fi 156, D-IAFZ, was modified by the Aerodynamische Versuchsanstalt
Göttingen for evaluating the so-called "boundary layer effect", where it carried out
numerous tests under the designation AF-2. Results of this research were absolutely
positive. The wings were altered during the evaluation program based on
recommendations of Dipl.Ing. Stüper von Ing. Grothy.

Fieseler Fi 256 and a Flettner Fl 282 helicopter. Both of these aircraft operated in the area around Berlin right up until the last few days of the war.

Fi 156 C of Nachtjagdgeschwader 200, which used this plane in 1945 as a liaison and cable laying aircraft. The terrain made the use of skis mandatory.

Above: Close-up shot of a Fi 156 C from NJG 200.
Above right: An Fi 156 C comes in for a landing.

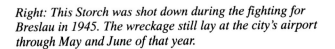

Right: This Storch was shot down during the fighting for Breslau in 1945. The wreckage still lay at the city's airport through May and June of that year.

Description

Airframe: Shoulder-wing with rearward folding wings, wings individually braced to lower fuselage longeron by paired V-struts. Struts braced to wings by N-support braces, fixed slats along the entire length of the wings, ailerons extending along half the wingspan and fitted with pressure overflow valves; landing flaps in wing section adjoining fuselage.

Enclosed cabin with three seats in tandem and windows to the sides, above and below, with luggage compartment behind seats. Tailfin fixed to fuselage, each tailplane braced from rear spar to tailfin and adjustable in flight, single-piece horn-balanced rudder. Twin spar wooden wings covered in fabric, slats made of dural, wood fabric-covered ailerons and landing flaps. Fuselage of tubular steel welded construction covered in fabric. Tailfin of tubular steel welded construction covered in fabric. All tail surfaces made of wood and covered with fabric. Two pairs of support braces join lower fuselage at same points. Coil spring and oil shock absorption. Landing impact absorbed by bottom fuselage longeron when extended, by wing attachment points when compressed. Wheels with hydraulic brakes. Tailskid fitted with coil spring and oil shock absorption.

Powerplant: Argus As 10 C, 240 hp, air-cooled. Fuel capacity: 150 liters. Fixed three-blade Heine airscrew.

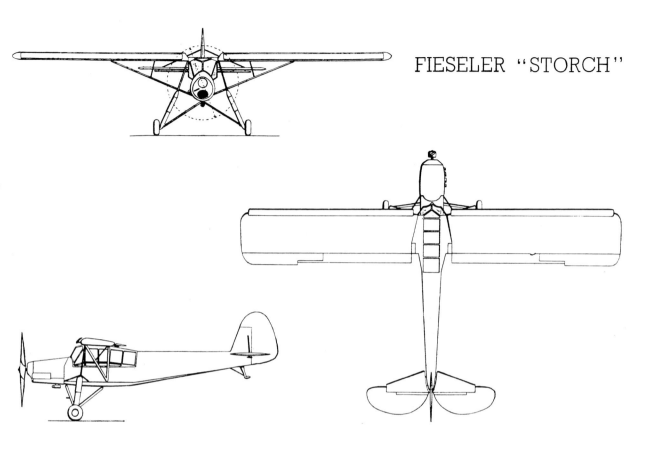

Three-view of the Fi 156.

Technical Data

Number of seats: 2 + one add'l as needed

Wingspan	
extended	14.25 m
folded	4.75 m
Height	2.80 m
Length	9.90 m
Weight (empty)	845 kg
Weight (loaded)	1320 kg
Range	400 kg

Max. speed	175 km/h
Landing speed	51 km/h
Rollout (w/brakes)	28 m
Landing rollout	
with 3.5 m/s	
headwind	47 m
Number, location and	
capacity of fuel	
and oil tanks:	
2 fuel tanks in wings, each	
with 75 liters	
total	150 liters
1 oil tank in fuselage	
ahead of firewall	15 liters

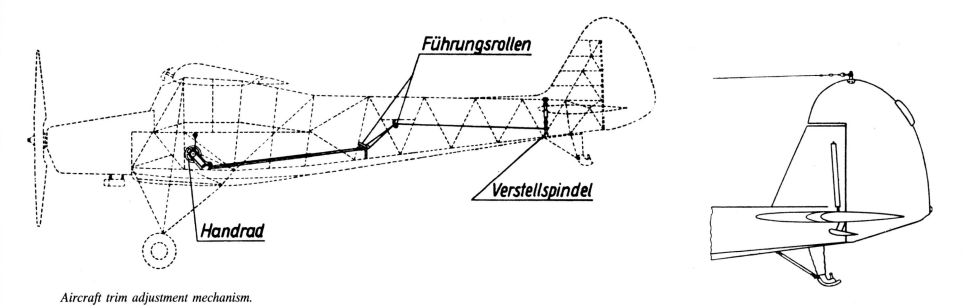

Führungsrollen

Verstellspindel

Handrad

Aircraft trim adjustment mechanism.

Rudder of the Storch showing radion antenna attachment point.

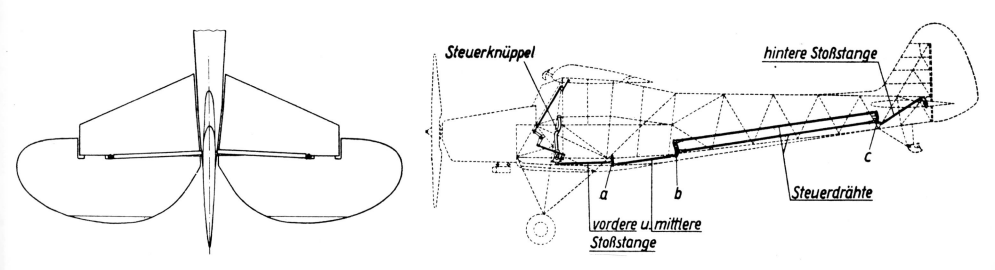

Top view of the horizontal stabilizer.

Steuerknüppel

hintere Stoßstange

a

b

c

vordere u. mittlere
Stoßstange

Steuerdrähte

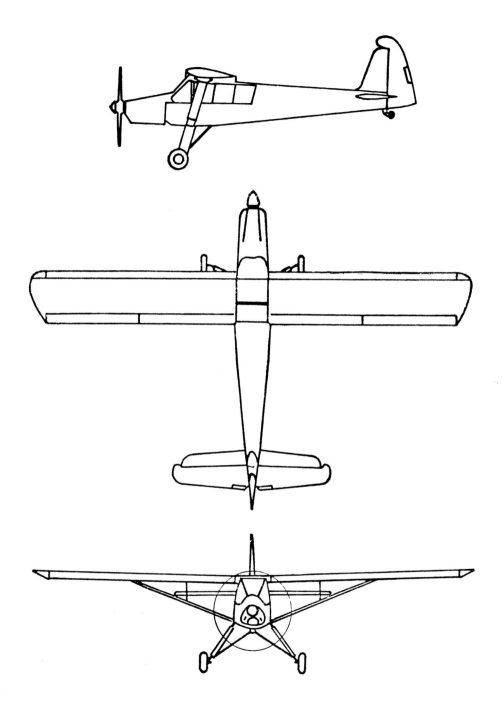

Above:
Crashlanding of a second Storch from NJG 200 in the early spring of 1945.

Below:
The Storch was flown by several countries both during and after the war. Here is a Storch in British markings following its capture in Italy.

Three-veiw of the Fieseler Fi 256.

Left:
A Morane-Saulnier Storch, which in its original configuration was supplied as the Morane-Saulnier MS 500.

Right:
Morane-Saulnier MS 502, a Storch with a 230 hp Salmson 9 AB engine, was delivered once supplies of the Argus As 10 C dried up.

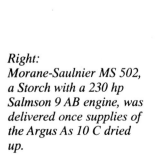

This German Storch was flown by the Polish Air Force for several years following the war. The tail surfaces to the right belong to a Polikarpov Po-2 supplied by the USSR.

In Spain the Storchs delivered before and during the war remained in both military and private use for a long time.

Right:
The Spanish Storchs
continued in service well
into the 1950s, thanks to a
large supply of spare
parts.

Left:
As in 1941/42
Libya, the
Storchs were well
suited to the
desert climate
prevailing in the
Spanish colonies
(here in
Morocco).

After the Czech firm of Mraz had built 64 Fi 156 Ds for the Germans, additional Fi 156 Ds were built after 1956 in the CSSR under the designation K. 65 "Cap" and flown by the Czech Air Force.

This well-preserved Czech-built Storch can be seen today in Prague's Museum of Technology.

The Ryan Company in San Diego, California, attempted to copy the Storch principle in 1940. Three prototypes of the Ryan YO-51 were built, but these never attained the performance of the Storch.

Left:
A Morane-Saulnier MS 500 Storch in service with the Thai Air Force.

In Japan the Nippon Kokusai Company built a modified Storch with the designation of Ki 76 (U.S. codename "Stella"). A Hitachi Amakaze Ha-42 280 hp engine served as the powerplant. Although the Ki 76 was an army aircraft, seven Ki 76s operated from the aircraft carrier Akitsu Maru in the anti-submarine warfare role.

HEINKEL He 100
World Record and Propaganda Aircraft

Hans-Peter Dabrowski

DORNIER DO 335 "PFEIL"

THE LAST AND BEST PISTON-ENGINE
FIGHTER OF THE LUFTWAFFE

THE FOCKE-WULF Fw 190
Fighters · Bombers · Ground Attack Aircraft

Heinz J. Nowarra

Junkers Ju 87

Ulrich Elfrath

JUNKERS Ju 52
HEINZ J. NOWARRA

MESSERSCHMITT Me 163 "Komet" Vol. II

M. Emmerling/J. Dressel

MESSERSCHMITT Me 262 Vol. II
The World's First Turbojet Fighter
Manfred Griehl

MESSERSCHMITT Bf 109
1936-1945

Heinz J. Nowarra

The Bf 109 F-trop of Hans-Joachim Marseille
of JG 27 (above) and the Bf 109 F of Adolf
Galland of JG 26 (foreground).

Ju 88
OVER ALL FRONTS

Joachim Stein